THE JOURNAL OF C

MW00655379

Vol. 1
(2012)

CFZ Press

Edited by Dr Karl Shuker
Typeset by Jonathan Downes/Sheri Myler,
Cover and Layout by SPiderKaT for CFZ Communications
Using Microsoft Word 2000, Microsoft Publisher 2000, Adobe Photoshop CS.

First published in Great Britain by CFZ Press

CFZ Press
Myrtle Cottage
Woolsery
Bideford
North Devon
EX39 5QR

ISBN: 978-1-905723-99-7

EDITORIAL

Welcome to the inaugural volume of the *Journal of Cryptozoology*. Following the demise of *Cryptozoology* (published by the now-defunct International Society of Cryptozoology), there has been no peer-reviewed scientific journal devoted to cryptozoology for quite some time. Consequently, the *Journal of Cryptozoology* has been launched to remedy this situation and fill a notable gap in the literature of cryptids and their investigation. For although some mainstream zoological journals are beginning to show slightly less reluctance than before to publish papers with a cryptozoological theme, it is still by no means an easy task for such papers to gain acceptance, and, as a result, potentially significant, serious contributions to the subject are not receiving the scientific attention that they deserve. Now, however, they have a journal of their own once again, and one that adheres to the same high standards for publication as mainstream zoological periodicals.

To that end, a distinguished peer review panel has been assembled (and has been greatly extended in recent months), whose members remain anonymous as is traditional and necessary for the independent evaluation of submissions to the journal, but which consists of some of the world's most eminent zoologists and associated researchers in their respective fields of technical expertise. And I was honoured to be invited by the journal's publisher, the CFZ Press, to become Editor - an invitation that I was delighted to accept.

It is planned that each volume of the *Journal of Cryptozoology* should contain three to four papers (depending upon their combined total length). These can be discussion or review articles concerning a given cryptozoological subject; research-related papers; or field reports. Full details concerning the required presentation style for submissions, preceded by an advisory section detailing important issues to consider when preparing a cryptozoological paper, can be found at the end of this journal.

All subscription enquiries should be addressed to the publisher at publisher@journalofcryptozoology.com

Down through the decades, cryptozoology has been defined in different ways by different researchers, with

3

some definitions much more restrictive than others. Consequently, it is important to make clear the definition – and therefore the scope of subjects available for papers – to which this journal adheres. For the purposes of relevance to this journal, a cryptid is a creature that is known to the local people sharing its domain (ethnoknown) but unrecognised by scientists. Retaining the categories included in pioneering cryptozoologist Dr Bernard Heuvelmans's cryptid-defining paper 'Annotated Checklist of Apparently Unknown Animals with which Cryptozoology is Concerned (*Cryptozoology*, vol. 5, pp. 1-26, 1986), such a creature may be any of the following:

1) A species or subspecies apparently unknown to science, including alleged prehistoric survivors (e.g. mokele-mbembe).

2) A species or subspecies presently unknown to science in the living state, but which is known to have existed in historical times and allegedly still persists today (e.g. thylacine).

3) A species or subspecies known to science but allegedly existing as a natural occurrence in a location outside its scientifically-recognised current geographical distribution (e.g. puma in the eastern USA).

4) A species or subspecies known to science but allegedly existing as an artificial occurrence (i.e. due to human intervention) in a location outside its scientifically-recognised geographical distribution (e.g. alien big cats in Britain).

5) An unrecognised non-taxonomic variant of a known species or subspecies (e.g. Fujian blue tiger; prior to its scientific recognition, the journal's logo creature, the king cheetah, was another example from this category).

In addition, papers dealing with fabulous, mythological beasts will be considered for publication in the journal if their subjects have direct relevance to cryptids (e.g. reviewing the similarity between a given lake monster from folklore and cryptids reported in that same lake in modern times).

Some cryptozoological researchers prefer to impose a lower size limit for cryptids arguing that a crucial aspect of a cryptid's definition is that it should be of unexpected form. However, as I have revealed time and again in my various books documenting new and rediscovered animals, some very notable, unexpected cryptids were also very small. This is exemplified by Kitti's hog-nosed bat *Craseonycteris thonglongyai*, scientifically described in 1974 but already known to the local Thai people, and so dramatically different from all other bats that it required the creation of an entirely new taxonomic family to accommodate it – yet it is no bigger in size than a bumblebee. Consequently, although this journal is primarily interested with 'classic' cryptids, i.e. those of large or relatively large size, whose apparent continuing existence undiscovered by science is therefore particularly surprising, papers dealing with interesting, unusual, or potentially significant cryptids of smaller size will also be considered for publication.

Conversely, unidentified animal-like (zooform) entities of an apparently paranormal nature, e.g. spectral Black Dogs, fall outside the scope of subjects with which this journal is concerned.

It is always exciting to be part of a major new development, and I believe that the *Journal of Cryptozoology* marks a major new development in the advancement and mainstream awareness of cryptozoology. I hope that you will too.

Dr Karl P.N. Shuker,
The Editor, *Journal of Cryptozoology*,
November 2012.

PUBLISHER'S NOTE

We have chosen the king cheetah as the emblem of the *Journal of Cryptozoology* as it is the perfect example of what cryptozoology *should* be concerned with. It is a particularly uncommon mutation of the cheetah which was first noted in Southern Rhodesia (Zimbabwe) in 1926, and was initially thought to be a separate species *Acinonyx rex*. Its status as a species was conclusively disproved nearly 60 years later when king cheetah cubs were born in captivity to parents with normal markings. Cryptozoology in its purest form is about the search for the truth concerning what Bernard Heuvelmans described in 1980 as 'unexpected animals'. The truth may not be as exciting, or even as newsworthy, as many pundits would like, but it is still the truth. And in the end, that is all that matters.

The *Journal of Cryptozoology* is published and funded by CFZ Press, which is owned by the Centre for Fortean Zoology. However, it is an entity completely separate from the CFZ and entirely independent of it. We have long been aware of the need for a peer-reviewed academic journal covering cryptozoology, and we feel that it is important that it is truly international in scope and independent of any pre-existing organisation or pressure group.

We think that this is a very exciting new venture, and are immensely proud to be involved as its publishers. We are sad to see that other publishing ventures have folded for lack of funding, and so CFZ Publishing Group will guarantee to maintain funding of this project for at least the first ten volumes, regardless of sales, subscriber figures, or other outside factors.

www.journalofcryptozoology.com

CONTENTS

A DIGITAL SEARCH ASSISTANT FOR CRYPTOZOOLOGICAL FIELD EXPEDITIONS

Andrew May
Crewkerne, Somerset, U.K.

ABSTRACT

A cryptozoological field expedition will typically have limited resources, and a limited amount of time, to search a large area for an elusive target. It is thus faced with the problem of how to use the available time and resources in the most efficient way possible. This situation is by no means unique to cryptozoology: similar problems are regularly faced by search and rescue teams, and by military forces where the object of the search may be anything from a submarine to a mobile missile launcher. Mathematical analysis of the search problem has led to the development of software tools designed to assist search personnel in a number of application areas. The aim of the present paper is to show how this could be extended to the field of cryptozoology. After a brief review of the relevant theory, the paper presents an illustrative example using a prototype 'Digital Search Assistant'. While the example is simplistic in comparison to a real-world situation, it is sufficient to illustrate the principles involved, and the potential benefits. The paper concludes with a consideration of how the software could be developed into a more user-friendly production version.

Keywords: Search; Software; Mathematical modelling; Statistics; Bayesian inference; Presence-absence

INTRODUCTION

In 1968, the United State Navy used a then-obscure mathematical technique called Bayesian search to locate the wreck of its nuclear submarine USS *Scorpion* (Sontag and Drew, 2000). The *Scorpion* was found using paper and pencil, but the same method was soon used as the basis for computerised search algorithms by both the US Navy and the US Coast Guard. Over the years, Bayesian search algorithms have increased in sophistication, and are now used in a range of applications from tactical decision aids for anti-submarine warfare (Bjorklund, 1990) to unmanned air vehicles for search and rescue missions (Goodrich *et al.*, 2008).

One of the reasons automated search algorithms have proved so useful is that they make full use of the 'negative information' collected during the search process, rather than simply discarding or ignoring it as an unaided human searcher might. If a search team expends several hours' effort in scouring an area without finding anything, there is a natural tendency to dismiss this as so much wasted effort. In the automated approach, however, it is treated as valuable 'negative information' about *where the target is not*, which through the use of mathematical formulae can be converted into information about *where the target is most likely to be*. Of course, this is merely good scientific practice — but professional search teams are not always trained in the scientific method.

A Bayesian 'Digital Search Assistant' provides guidance to the user on the allocation of resources to the most likely target locations, and dynamically updates this guidance as new information — either positive or negative — is obtained. The relevance of such a tool to cryptozoology is clear. A typical cryptozoological field expedition has many features in common with the sorts of situation where the Bayesian approach has already proved successful: a large search area, limited resources, a limited amount of time, a low probability of detecting the desired target, and a large quantity of negative information.

The novelty of the present study lies in the application of the particular technique — Bayesian search — to the particular field of study — cryptozoology. Bayesian techniques of one kind or another are well established in other areas of the biological sciences, while the use of mathematical statistics in cryptozoology is not a new idea in itself. Methods based on the binomial distribution have been developed to estimate the probability of existence of cryptic species in the absence of positive observations (Guynn *et al.*, 1985; Reed, 1996). Mathematical models have also been used to place limits on the present population size of species generally presumed to be extinct (Payne, 2010). Studies such as these indicate how rare organisms may exist at sufficiently low densities that they can successfully evade detection using conventional techniques, which *inter alia* reinforces the rationale for the model proposed in the present paper.

Although the paper focuses on the application to cryptozoology, a similar approach could be applied to the tracking and monitoring of rare but known species, and hence have wide implications in the fields of conservation and ecology.

The next section briefly summarises the theory of Bayesian search. Readers who are no interested in the technical details can skip over this and go straight to the section following in which discusses how the technique might be applied to cryptozoology. This is followed by worked example, using prototype software, to illustrate the principles involved and th potential benefits of the approach. The paper concludes with a discussion of how the prototyp might be further developed into a useable field tool.

BAYESIAN SEARCH THEORY

Bayesian search is a subset of the broad field of Bayesian inference. This is one of the tw major schools of statistical inference, the other being the frequentist school. Both schoo work with hypotheses and probabilities, but handle them in subtly different ways. Th

Bayesian approach is adopted here because its mathematical formulation is more convenient when a large number of hypotheses need to be evaluated recursively over a period of time.

The use of Bayesian inference in the evaluation of anomalous phenomena has been proposed in the past (Sturrock, 1994), and it is well-established in the field of ecology (Ellison, 2004). Much of the work in the latter field is concerned with presence-absence and occupancy modelling, and in the efficient allocation of resources to the monitoring of rare species where the number of observations is small but non-zero (Yoccoz et al., 2001; Conroy et al., 2008). Perhaps most relevant to cryptozoology is the use of Bayesian inference to estimate presence when a species is undetected (Peterson and Bayley, 2004).

The essence of the Bayesian approach is that it assigns *probabilities* to *hypotheses*. In the case of Bayesian search, a **hypothesis** relates to a possible current state of the target. In the case of a stationary target, the state is simply defined by its location on a map — in other words, by two numbers such as northing and easting. If the target is moving, then its state will be more complex, requiring several more numbers to characterise its motion and behaviour. At the start of the search, there will be a large number of possible hypotheses as to the state of the target — typically many hundreds or even thousands. We will denote the number of hypotheses by N, and the ith hypothesis ($i = [1, N]$) by H_i.

Each hypothesis H_i is characterized by a **probability** $p(H_i)$ that it is true, where p is a number between 0 (if the hypothesis is definitely false) and 1 (if the hypothesis is definitely true). The probability that hypothesis H_i is false is $1 - p(H_i)$. At least one of the hypotheses must be true, so the probabilities over all the hypotheses add up to 1:

$$\sum_{i=1}^{N} p(H_i) = 1 \quad (1)$$

The probabilities associated with the various hypotheses evolve over time as new information is acquired. Before the search starts, the probabilities are assigned initial values based on the best available information. These initial values are called **prior probabilities** and can be denoted by $p_0(H_i)$. If the search is broken down into a series of discrete steps, the probability after the completion of the nth step can be denoted by $p_n(H_i)$.

The result of the nth search step will be the acquisition of some new information. This may be positive information, for example if traces of the target are detected, but more often it will be negative information — in other words, nothing is detected. Both types of result, positive and negative, are treated in the same way. Let the result of the nth search step be denoted by R_n. Then the question to ask, for each hypothesis H_i, is: "What would be the probability of obtaining result R_n, assuming that hypothesis H_i is true?" The answer to this question can be denoted by $p(R_n \mid H_i)$, and is called the **likelihood**.

As with the prior probabilities, the likelihood values are inputs to the process. Suitable values for priors and likelihoods should be developed in consultation with relevant domain experts. In an uncertain field like cryptozoology, there may be considerable guesswork and intuition involved at this stage, so it is important to record all the underlying assumptions. As with any mathematical model, the output will be garbage if the input is garbage!

The likelihood is the probability that result R_n would be observed given that hypothesis H_i is true. But what we really want is the inverse of this: the probability that hypothesis H_i is true given that result R_n has been obtained. Fortunately, there is a mathematical formula called **Bayes' Theorem** (Chapman, 1981) that performs the required inversion:

$$p_n(H_i) = p(H_i \mid R_n) = p_{n-1}(H_i) \times p(R_n \mid H_i) / p(R_n) \qquad (2)$$

The term $p(R_n)$ on the right-hand side is simply the sum of $p(R_n \mid H_i)$ over all N hypotheses, and is required to ensure that Equation (1) is still obeyed. Although the equations may look daunting, they can easily be evaluated in a few lines of computer code. Equation (2) is recursive: it updates the probability distribution from step $n-1$ to step n. This is a very powerful result: it means that our current probability distribution always embodies every single piece of information we have about the situation: our preconceived notions embodied in the prior probability, any positive information that has been detected, and — just as important — negative information about where the target is not.

APPLICATION TO CRYPTOZOOLOGY

The sophisticated search tools developed for military applications are generally aimed at mobile targets, and these involve quite complex mathematics. The case of a stationary target is much simpler, and is all that will be considered here. While most cryptids are of course mobile, there are nevertheless situations where the 'target' of interest is effectively stationary:

- A colony of small, slow moving animals, such as worms, arthropods or amphibians, which is effectively static from the point of view of a human searcher.

- A large, social cryptid such as a primate, where individuals might roam quite widely, but return to a localised den or nest which would then be the target of the search.

- The skeletal, mummified or fossil remains of a cryptid reputed to have dwelt in a particular area.

In the case of a stationary target, a 'hypothesis' is characterised by the target's location on a map. The simplest approach is to divide this map into a grid of cells, each of which corresponds to an area that could be searched in a specific length of time such as a day. In the terminology of the previous section, this unit of time would correspond to a search 'step'. The total number of cells would be $N-1$. Why $N-1$ and not N? N is the total number of hypotheses, and the probabilities of all N hypotheses must add to 1 (Equation 1). There are $N-1$ hypotheses corresponding to the target being in each of the $N-1$ cells, and then one final hypothesis - called the **null hypothesis** — that the target is not in any of the cells.

The search effort should be focused on whichever cell currently has the highest probability. When a search step is completed, the probability map should be updated using Equation 2, and the next step directed at whichever cell now has the highest probability. Depending on the circumstances, this may be the same cell as the previous step, or an adjacent cell, or a completely different cell.

The output of the Digital Search Assistant is advisory only; the search leader is free to use his or her discretion. For example, if the tool suggests searching a cell twenty kilometres away, and there is another cell two kilometres away that has almost as high a probability, then it would be sensible to search that one first. But whichever search is undertaken, the results must be entered into the Search Assistant, or the probability map will cease to be accurate.

The Search Assistant requires two types of input data, known in Bayesian jargon as 'likelihoods' and 'prior probabilities'. Both of these can be drawn from expert knowledge. The likelihoods, for example, are essentially the answers to the following questions:

- If the target is in a particular grid cell, what is the probability (on a scale of 0 to 1) that positive evidence of it will be found in that cell after a day's search?
- If the target is in a particular cell, what is the probability that positive evidence will be found in an adjacent cell after a day's search?
- If the target is in a particular cell, what is the probability that positive evidence will be found in a non-adjacent cell after a day's search?

There is no need to ask the same three questions for each of the $N-1$ cells, since the answers (in probability terms) are assumed to be the same for all cells. Equally, there is no need to ask the corresponding questions about negative information, since the probability of obtaining negative information is simply 1 minus the probability of obtaining positive information. It is possible, however, that *different types* of positive information (e.g. footprints, evidence of feeding, tentative sightings, etc) will be associated with different probabilities, in which case they will require a separate set of questions.

With regard to establishing the prior probability distribution, one approach would be as follows. For each of the $N-1$ cells, consider a number of factors, such as proximity to food source, proximity to drinkable water, presence of shelter, lack of predators and lack of natural hazards. Suppose for the sake of argument we come up with ten such factors. If each 'tick in the box' accrues a score of one, then the result is a score somewhere between 0 and 10 for each cell. These scores can then be converted into probabilities by rescaling them ('normalising' them, in mathematical jargon) such that the result obeys Equation (1).

If a cell is initialised with zero probability, then it will always remain at zero probability (see Equation (2): zero multiplied by anything else is still zero). For this reason, a prior probability of zero should only be assigned to cells where it is physically impossible for the target to be for example, a large expanse of water, if the target is a land-dwelling creature). The 'null hypothesis' should be initialised with a small but non-zero probability — this will be found to rise automatically as the search progresses without the target being found.

AN ILLUSTRATIVE EXAMPLE

A prototype 'Search Assistant' tool was produced using Microsoft Excel with embedded Visual Basic macros, and tested on a fictitious example scenario. Both the software and the example scenario were kept as simple as possible consistent with demonstrating the principles involved and the potential benefits of the approach.

The scenario is based on the map shown in Fig. 1, which shows the search area divided into a 23×15 grid of square cells. It is assumed that at most one of these cells contains the stationary target that is being searched for. The nature of the target is not really important, but it might for example be a 'Bigfoot den'.

Fig. 2 shows the prior probability distribution, based on a 'common-sense' assessment of the terrain features. For example, it is assumed that the probability that the target is close to a town or road is low, whereas the probability that it is in an area entirely covered by water is zero. On the other hand, the probability that it is relatively close to a source of fresh water is high.

When a cell is searched, the probability associated with that cell and all other cells is updated using Equation (2), with different likelihood values depending on whether or not the search turns up positive evidence. As with the priors, the likelihoods are user inputs based on a mixture of common sense and 'gut feeling'.

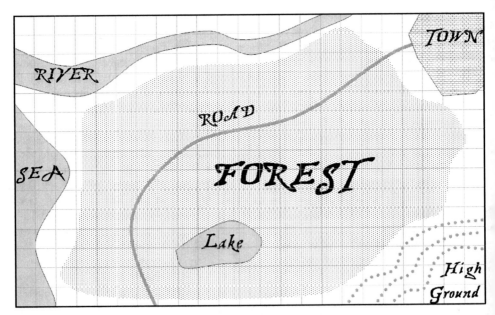

Fig. 1: Map of the Example Scenario

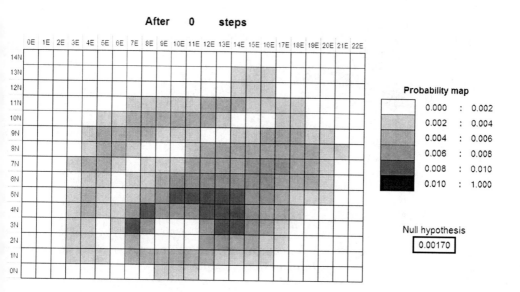

Fig. 2: Probability map prior to start of search

The probability map will change as the search progresses, but the precise way that it changes depends on the information obtained. Figs 3 and 4 show two possible configurations after three search steps. In both cases the three steps were identical, involving a search of cells 4N-13E, 5N-10E and 4N-8E in that sequence.

In the first case (Fig. 3) no evidence of the target was found at any step, whereas in the second case (Fig. 4) positive evidence, e.g. a footprint, was found at the third step.

Fig. 3 illustrates the cumulative effect of 'negative information'. The probability map looks quite different from the 'prior' distribution of Fig. 2, with the darkest cells — the area where the search should now be focused — shifted to the north-east. Fig. 4, on the other hand, illustrates the effect of positive information.

Again it looks very different, both from Fig. 2 and from Fig. 3. There is a concentrated area of high probability around cell 4N-8E, where the positive evidence was found, while further afield the colours have all become much lighter. This suggests that the next few search steps should be focused in this localised region.

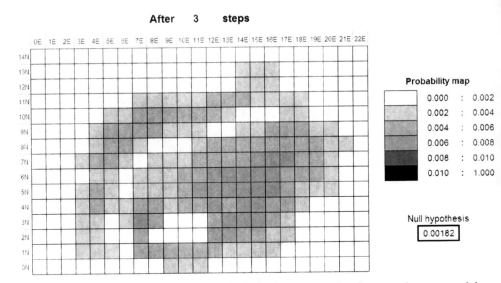

Fig. 3: Probability map after three cells have been searched, assuming no positive information was found

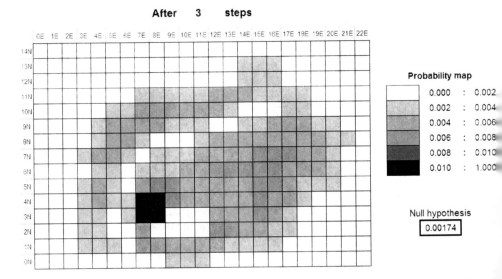

Fig. 4: Probability map after three cells have been searched, assuming positive information was found in one of the cells

CONCLUSIONS

This paper has argued that a Digital Search Assistant, of the kind used in various military contexts and in search and rescue operations, could be a useful new tool for cryptozoological field expeditions. A simple example has been used to illustrate how such a tool might function, and how it could be used to direct limited search resources to the most probable locations. Use of Bayesian search theory ensures that all the available information — negative as well as positive — is employed in building up the probability map.

The prototype described here consists of an Excel workbook containing four worksheets and approximately 120 lines of Visual Basic code. To be useable on field expeditions, this could easily be developed into a laptop-based application with a more 'user-friendly' input/output interface. However, a more practical implementation might consist of a mobile phone or tablet application, particularly if GPS was employed to record the searcher's current position automatically. It would also be possible to network applications on several devices, allowing multiple search teams to benefit from information collected by other teams as well as themselves.

REFERENCES

- Bjorklund, B.R. (1990). *Probabilistic Observations on Antisubmarine Warfare Tactical Decision Aid.* United States Naval Postgraduate School (Monterey).
- Chapman, M. (1981). *Decision Analysis.* Her Majesty's Stationary Office (London).
- Conroy, M.J., Runge, J.P. *et al.* (2008). Efficient estimation of abundance for patchily distributed populations via two-phase, adaptive sampling. *Ecology,* 89(12): 3362–3370.
- Ellison, A.M. (2004). Bayesian inference in ecology. *Ecology Letters,* 7: 509–520.
- Goodrich, M.A., Morse, B.S. *et al.* (2008). Supporting wilderness search and rescue using a camera-equipped mini UAV. *Journal of Field Robotics,* 25(1): 89–110.
- Guynn, D.C., Downing, R.L. and Askew, G.R. (1985). Estimating the probability of non-detection of low density populations. *Cryptozoology,* 4: 55–60.
- Payne, C. (2010). The continuing strange tale of the thylacine. *The Anomalist,* 14: 125.
- Peterson, J.T. and Bayley, P.B. (2004) A Bayesian approach to estimating presence when a species is undetected. *In:* Thompson W.L. (ed.), *Sampling Rare or Elusive Species.* Island Press (Washington D.C.).
- Reed, J.M. (1996). Using statistical probability to increase confidence of inferring species extinction. *Conservation Biology,* 10(4): 1283–1285.
- Sontag, S. and Drew, C. (2000). *Blind Man's Bluff.* Arrow Books (London).
- Sturrock, P.A. (1994). Applied scientific inference. *Journal of Scientific Exploration,* 8 (4): 491–508.
- Yoccoz N.G., Nichols J.D. and Boulinier, T. (2001). Monitoring of biological diversity in space and time. *Trends in Ecology and Evolution,* 16(8): 446–453.

THE QUEENSLAND TIGER: FURTHER EVIDENCE ON THE 1871 FOOTPRINT

Malcolm Smith
10 Carl Place, Bracken Ridge,
Queensland 4017, Australia.

ABSTRACT

The alleged marsupial 'tiger' of north Queensland has been the subject of infrequent reports of sightings, most of which are very old. The best evidence remains a footprint recorded in 1871 by surveyor Alfred Hull, and published by the Zoological Society of London. New material is now provided in the form of a contemporary newspaper account by the witness. It includes more details of the date and site and indicates that the footprint, and hence the animal, was twice as large as originally believed.

INTRODUCTION

Although never officially described, the legendary 'Queensland Tiger' once held a semi-official position in the Australian fauna because, for the first half of the 20th century, it was listed in the two popular guidebooks on Australian mammals (Le Souef and Burrell, 1926; Troughton, 1941 and all subsequent editions). According to legend, it is a large, striped, cat-like animal, presumed to be a marsupial, inhabiting the dense forests of northern Queensland, but also reported from isolated southern areas of the state. Since the publication of the above-mentioned guide books, occasional reports have drifted in (Heuvelmans, 1958; Healy and Cropper, 1994; Shuker, 1995; Smith, 1996). However, the 'tiger' is seldom reported these days - either because it is genuinely rare or non-existent, or because it has slipped out of the collective consciousness.

In any case, the best concrete evidence for its existence remains a footprint recorded from north Queensland in 1871, as it cannot be assigned to any known species. The original report was in a letter by Walter T. Scott of Cardwell, Queensland, dated 4 December 1871, and submitted to the Zoological Society of London (Scott, 1872).

A Mr. Hull, Licensed Surveyor, was lately at work with a party of five men, surveying on the Murray and Mackay rivers, north of Cardwell. They were lying in their tents one night between eight and nine o'clock, when they were all startled by a loud roar close to the tents. They seized their guns and carefully reconnoitred; but the animal had departed. In the morning they found the tracks of the unknown visitor, of which Mr. Hull took the measurements and a rough sketch. I send you part of a leaf of Mr. Hull's field-book, containing the original sketch - and also his drawing of the track, of the natural size. Mr. Hull assures me that the drawing was a very faithful one, the soft ground having taken the impression with all its details. I have also examined some of the men who were with Mr. Hull. They all tell the same story, and say they heard the animal three nights in succession.

The connection between the original sketch and the drawing of the track is not at all clear – nor is what was meant by "the natural size". However, the drawing which the journal did publish (Fig. 1) was labelled "Footprint of 'Native Tiger', reduced by half". As published in the journal, it was 1.2 inches [3.1 cm] long, including the claws, and 1.25 inches [3.2 cm] wide. Doubling that would still make the animal rather small – comparable to (say) a blue cattle dog. There the matter has rested until now.

Fig. 1: Hull's original published drawing of the footprint.

NEW MATERIAL

In 1996 I was able to contact William Kitson of the Museum of Mapping and Surveying, Brisbane, who was preparing a book on Alfred Hull's sketches and travels. He told me that the sketch does not occur in the field-book, and the author probably meant his diary, which was not in his (Kitson's) possession. However, he did point out that Hull serialised his exploration in the newspaper, the *Brisbane Courier* under the name of "Taff of Tolosa" (Tolosa being the name of his property).

RESULTS

The relevant entries, taken from the 17 October 1871 issue of this newspaper, are as follows:

- 24th. [August] - Left camp this morning in the boat, and rowed up the Murray to the Bellenden Plains landing place, about twenty-five miles up the river . . .
- 26th. - Shifted camp lower down the river, and were disturbed in the night by the alligators [i.e. crocodiles] bellowing, and a native tiger roaring close to our camp; the tiger came within one hundred yards of our tent, but we could not get a shot at him, owing to the darkness and the scrub. I believe it is not generally known that there is such a thing as a native tiger in Queensland, but it is nevertheless a fact.
- 27th and 28th. - Working down the river; scrubs very dense and swamps deep, and lined with cutting rushes from nine to ten feet high. Where the country is dry, which is very rarely the case, on this (the south) bank the soil is poor and sandy; on the north side, from the Bluff upwards, the country is far better, being dry and open, lightly timbered, and well grassed. Saw the tracks of the tiger and measured them - four inches long by four and a-half inches wide - so that there is no doubt about the existence of a very large animal of the cat tribe in these scrubs.

DISCUSSION

From this, the following conclusions can be reached:

(i)   The event took place in the evening of 26 August 1871, and the tracks were found an unstated distance away the next morning. The connection between the roaring and the tracks is assumed, but not proved. There is no mention of roaring three nights in a row, but it is possible that on the previous nights it was not close enough to be worth recording.

The reference to roaring is curious. I know of no other case where it has been reported in connection with the alleged Queensland tiger. Indeed, it is not normally used of any native animal, although the bellowing of crocodiles or koalas, or the booming call of the cassowary might occasionally be given that name. However, Hull clearly distinguished it from the sound made by a crocodile, and the dense, damp undergrowth he described would be unlikely to harbour koalas. In any case, the "roaring" must have been alarming enough to bring them out of their camp at night, and the juxtaposition of an unidentified sound with an unidentified footprint some distance away would be a significant coincidence if they were unrelated.

(ii)   The site was on the south bank of the relatively small Murray River, less than 25 miles from its mouth. This would be about 20 miles northwest of Cardwell, at about 18°S, 145°50' E.

(jj)   Most importantly, the drawing had been "reduced one half" twice - once by Hull, and once again by Scott or the editor. The footprint was, in fact, large enough to belong to an animal the size of a large dog or small leopard - consistent with the roaring heard, and the testimony of tradition.

Fig. 2: General outlines of footprints of (a) macropod (top left), where two feet are placed together; (b) canid (top right); (c) felid (bottom left); (d) thylacine (bottom right). The front prints are shown. Although the rear prints differ slightly – and, of course, differ according to the precise species – this is of no significance in comparison to the Hull footprint.

What could have made it? It is certainly consistent with a mammalian predator. However, th combination of a small, ovoid pad surrounded by four large toes of equal size, in more or les the same plane, without a space between pad and toes, is not obviously referrable to any known species. Hoofed domesticated animals are excluded, and the only native herbivores in the required range are macropods: kangaroos and wallabies. Their feet are designed for hopping, and are elongated along the axis of the fourth toe, with a smaller fifth toe to the side, while toes 2 and 3 are syndactylous, i.e. fused together as a very small grooming implement. However, occasionally, a macropod will place its two hind feet together. If only the distal pads come in contact with the ground, this can lead to a confusing composite print such as that in Fig. 2(a), with the small syndactylous toes obscured. The result is the appearance of a single foot with a deeply divided two-lobed pad, and four toes of unequal size. This is nothing like the Hull footprint.

The same can be said for known large predators. A typical canid footprint is shown in Fig. 2 (b): a subtriangular pad with a smooth, concave rear, and the outer toes more recessed than the middle ones. For decades there have been reports of alien big cats – usually labelled pumas or black panthers – in Australia. However, fresh light was cast on the mystery when a huge black cat was shot by Kurt Engels in 2005 (Williams and Lang, 2010), because it was revealed by both hair analysis and DNA to belong to the domestic cat *Felis catus*. There can no longer be any doubt that feral domestic cats in the continent have grown to the size of large dogs, or even small leopards. While this may well confuse the picture today, it is unlikely that they were present in 1871. In any case, a typical felid footprint resembles that in Fig. 2(c): the pad is truncated in front, and bears three lobes at the rear. Normally, the retracted claws leave no mark, but they may show in the sort of soft condition where the Hull footprint was found. Finally, the footprint of the thylacine *Thylacinus cynocephalus*, historically restricted to Tasmania, bore a pad with three lobes in the rear, and a triangular extension in front, as per Fig. 2(d). The front paw also often displayed a small fifth toe.

If known animals can be ruled out, what about prehistoric ones? Heuvelmans (1958) nominated the marsupial lion *Thylacoleo carnifex*, and most non-specialist commentators have followed suit. However, there remain problems. *Thylacoleo*'s hind foot had two large toes and two smaller, syndactylous toes, while the forepaw bore four toes, more or less equal in size, and a smaller, semi-opposable pollex to the side (Wells and Nichols, 1977). Thus, it could be consistent with the Hull footprint only if the pollex failed to imprint, or was overlooked. This is unlikely if there were multiple footprints, and Hull's narrative referred to "tracks", rather than "a track".

We are thus left with the conclusion that the footprint represented a true unknown.

It should also be noted that, as mentioned in an earlier publication (Smith, 1996), I have seen a representation of a very similar footprint, but with one extra toe, more than 500 miles to the south. This is on a cliff face in the Carnarvon Gorge National Park, at approximately 25°5' S, 148°10'E. It is in the midst of a large number of other Aboriginal petroglyphs of the footprints of more common species, and a sign erected by the park authorities points out, correctly, that it is not referable to any known species. The extra toe is the same size as the others, and in the same plane – which is also not easily referrable to *Thylacoleo*. This would appear to be confirmatory evidence, both that Hull's drawing is accurate, and that the unknown species was – and perhaps still is – widely distributed.

ACKNOWLEDGEMENTS

I would like to express my appreciation to William Kitson for this information, and to two distant relatives of Alfred Hull, Lucille Andell and my cousin, Christine Downie, who directed me to him.

REFERENCES

- Healy, T. and Cropper, P. (1994). *Out of the Shadows: Mystery Animals of Australia.* Pan Macmillan (Sydney).
- Heuvelmans, B. (1958). *On the Track of Unknown Animals.* Rupert Hart-Davis (London).
- Le Souef, A.S. and Burrell, H. (1926). *The Wild Animals of Australasia: Embracing the Mammalogy of New Guinea and Nearer Pacific Islands.* George C. Harrop (London).
- Scott, W.T. (1872). Letter addressed to the Secretary concerning the supposed "Native Tiger" of Queensland. *Proceedings of the Zoological Society of London*, 1872, p 355.
- Shuker, K.P.N. (1995). *In Search of Prehistoric Survivors: Do Giant 'Extinct' Creatures Still Exist?* Blandford Press (London).
- Smith, M. (1996). *Bunyips and Bigfoots: In Search of Australia's Mystery Animals.* E. J. Dwyer (Sydney).
- Troughton, E.L-G. (1941). *Furred Animals of Australia.* Angus and Robertson (Sydney).
- Wells, R.T. and Nichols, B. (1977). On the manus and pes of *Thylacoleo carnifex* Owen (Marsupialia). *Transactions of the Royal Society of South Australia*, 101: 139-146.
- Williams, M. and Lang, R. (2010). *Australian Big Cats: An Unnatural History of Panthers.* Strange Nation (Hazelbrook).

HUNDA 'SCAPASAURUS' PHOTO (RE)DISCOVERED, WITH EXPLANATIONS OF DESCRIPTIVE TRENDS IN RELATION TO PSEUDOPLESIOSAURS

Markus Hemmler
Königsberger Str. 12, 75417 Mühlacker, Germany.
Email: Markus.Hemmler@gmx.de

ABSTRACT

Between the years 1941-1942, two so-called 'pseudoplesiosaurs', actually washed up basking shark *Cetorhinus maximus* carcases, were found on the Orkney Islands, Scotland, United Kingdom. An assumed photograph of the second carcase from the island of Hunda was recently discovered in the collections of the Orkney Library and Archive. To verify and document this find together with its apparent historical context, the records of 'Scapasaurus #2' in the literature will be presented. Based upon this photo, a second example of the common descriptive trend for using the terms 'antennae', 'feelers', 'finger-like projections', or 'horns' in relation to pseudoplesiosaurs will be revealed. Also the reported 'blow holes' or 'holes' in the head will be revealed as a further descriptive trend regarding such carcases, and the morphological background explained.

INTRODUCTION

Some 'sea monster' carcases seem on first sight to correspond in superficial external shape to an extinct marine plesiosaur, often appearing to possess a small head, a long thin neck, a large body with fins, and a pointed tail. In all cases where sufficient material or data for identification was present, however, they turned out to be the carcase of a basking shark *Cetorhinus maximus* (Fig. 1).

Fig. 1: From basking shark to pseudoplesiosaur (Markus Bühler)

Eventually, this led to the coining of the term 'pseudoplesiosaur' (Cohen, 1989). Within just few weeks around 1941-42, two pseudoplesiosaurs were discovered and reported from th Orkney Islands, Scotland, United Kingdom.

In August 2009, the author found in an article from the German newspaper *Bild* ('Was i dranan "Nessi", "Yeti" & Co?') a photograph (Fig. 2) of the first carcase that washed up Deepdale Holm (Fig. 3), located on Mainland, Orkney ('Fresh light on the mystery of th

Holm shore "Monster"'; 'Orkney Monster - Artist's impression'; 'The Orkney Monster - First pictures'). It is most probably one of 16 official photos that were prepared for the authorities by William S. Thomson, of the King Street Studio, Kirkwall, for scientific purposes and general press use ('More about the 'Monster''; 'Basking Shark or Scapasaurus?').

Fig. 2: Photograph of the Deepdale Holm 'Scapasaurus' (William S. Thomson)

Emphasising the similarity to prehistoric sea animals and commemorating the region Scapa Flow where it was found, this carcase was christened by the investigating naturalist and author James Marwick, Provost of Stromness, with the name '*Scapasaurus*' (quoted in 'Fresh light on the mystery of the Holm shore 'Monster'').

When reviewed in 1978, however, this name was exposed as a nomen nudum, because it does not meet the requirements of the International Code of Zoological Nomenclature (Bland and Swinney, 1978).

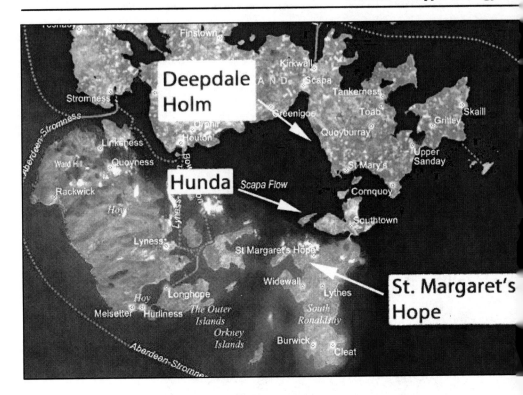

Fig 3: Map of Scapa Flow with relevant locations.
(http://commons.wikimedia.org/wiki/File:Orkney_Islands.jpg)

Within two weeks of the first find, another was discovered. 'Scapasaurus #2' was reported from the island of Hunda ('Second 'Monster' located'). An analysis had previously been published in the literature ('More about the 'Monsters''; 'St Margaret's Monster'; Dinsdale 1966); however, no picture was known to exist. As the author's correspondent Claire Jaycock recently discovered a hitherto unknown picture apparently showing the Hunda carcase, this evidence will now be examined and documented in context with possibly the most complete history of 'Scapasaurus #2' in a single publication.

MATERIAL AND METHODS

Articles from 1942 in the Orkney newspapers *The Orcadian* and the *Orkney Blast* as well as photograph of the second (Hunda) carcase were obtained from the Orkney Library and Archive. Morphological comparisons, explanations, and terms follow Izawa and Shibat (1993), Fairfax (1998), Hamlett (1999), and de Lulius *et. al.* (2007). Additional information regarding cases of pseudoplesiosaurs have been obtained from sources archived in the collection of the author and will be listed in the results.

To ensure historical authenticity, the story of the Hunda carcase has been cited solely from Orkney newspapers; and to ensure historical completeness, the whole year of *The Orcadian* (Claire Jaycock, pers. comm., 24 July 2011) and the *Orkney Blast* from January to May 1942 (Sarah Maclean, pers. comm., 20 April 2012) has been searched and double-checked with Dinsdale (1966). To determine the locality, the harbour master of St Margaret's Hope, Nick Lovick, was contacted. The photograph of the skeleton was carefully examined and its morphological features compared with the above-mentioned references.

RESULTS

In a letter to the local newspaper *The Orcadian*, W. Campbell Brodie of St Margaret's Hope (Fig. 3) gave a detailed first-hand account about this second carcase (quoted in 'Second 'Monster' located'):

A SECOND "SCAPASAURUS"?

Less than a fortnight after the appearance on the beach at Deepdale, Holm, reports reached me that a similar creature had been washed ashore on the Hope of Hunda, also in the Scapa Flow area.

Unfortunately several weeks had elapsed between the time of its first being observed and that of my setting out to investigate.

However, on February 5th, along with James Mcdonald and Andrew Laughton I made the journey across the island of Burray and thence to the island of Hunda.

We were rather sceptical of the reported dimensions of the second Orkney "Monster" and we intended to verify these for ourselves.

...On arrival at the very stony beach on the island a strange sight met our eyes. A huge elongated yellowish-coloured creature lay embedded in the sea-wrack, wedged firmly between the boulders.

The Skull.

Closer inspection revealed that this was indeed no ordinary denizen of the deep, but truly a creature of "Monster" proportions. All that remained was the skull and vertebral column, together with several appendages of fins consisting of cartilaginous material.

There were suggestions of a hump of a fatty nature, on the top of which was another cartilaginous fin, similar in structure to the larger fins lying beneath the vertebral column. The back of the latter was covered with greyish-black tough skin with a hairy appearance.

From some parts we obtained hair of a coconut fibre nature, about four to six inches [10-15 cm] in length.

The skull of the creature was composed of a gristly substance, definitely not bony and there was no evidence of teeth. On the anterior part were two antennae four inches long, and laterally two large sockets which may have contained the eyes, could be observed.

Length of 28 Feet [8.5 m].

On a raised part of the skull was a large cavity, possibly a blower hole, about midway along its dorsal surface. Immediately behind this were two smaller holes, all communicating with the mouth region.

Of a mouth or lower jaw there was no sign although we found a cartilaginous part, somewhat like a horse saddle in shape. We could not discover the articulations for this part or any part of the skeleton.

We proceeded to measure the carcase. From the tips of the antennae to the two rudder-like appendages in the tail region, the length was 28 feet [8.5 m] approximately, there being 65 vertebrae making up the spinal column.

The head alone measured two feet [60 cm] long and one foot [30 cm] broad at its widest part.

The appearance of the vertebrae was unusual, there being no spinous processes present. Each individual vertebra was bamboo-like in appearance and was joined to the preceding one with an elastic-like membrane.

Skeleton Salved.

One of the larger flipper parts measured three feet in length [90 cm] and consisted of 17 cartilaginous appendages.

The only sign of alimentary tract was a stomach-like part made up of gill-like structures. This may have been like the swim bladder found in ganoids.

We were convinced that the creature was of a similar species to that found at Deepdale, Holm, that we hired a boat from Mr John McBeath, a local hirer, and removed the skeleton in sections to St. Margaret's Hope, and now we are exhibiting it locally in aid of the Red Cross funds.

Campbell Brodie once again wrote a letter to the *Orcadian* on 17 February 1942 (Brodie 1942):

Sir. – Regarding the second Orkney "monster" located on the island of Hunda, the amount raised by voluntary collection in this little village, in little more than a week, was £6, and a cheque for that amount has been forwarded to Mr. W. J. Heddle, Kirkwall, for Red Cross purposes.

Regarding the "monster" itself, the local inhabitants are very interested in the question of its identification.

I have talked with probably 50 per cent of the fishermen at present engaged around the Scapa Flow area and I have still to meet one who believes the "monster" to have been either a shark or a whale. One trawler skipper compared it with a creature, long and slender in appearance and covered with long, brownish hair, which, he stated, rose above the water and attacked the ship's mizzen sail. That "monster" had a very long neck and a head resembling that of a cow. This happened near Hoy.

Mr David Wylie of Burray, also tells me that a creature of similar appearance was seen in Watersound between Hurray and South Ronaldsay a little more than a year ago.

So there is just the possibility that our "monster" here, besides aiding the war effort, may prove to be of scientific value, even new.

I intend to preserve the head and neck part and hope to dispatch it south for examination in the immediate future.

One thing is certain, we in South Ronaldsay are very proud of our "monster" – Yours etc.

W. CAMPBELL BRODIE

On 19 February 1942, *The Orcadian* published a letter about the Hunda carcase's identification as a basking shark by Dr Stephen Keeper (then curator of the Natural History Department of the Royal Scottish Museum in Edinburgh), as well as a letter to the editor from two scientifically-educated soldiers identifying the Hunda Beast as a basking shark too quoted in 'More about the 'Monsters''). Just a day later on 20 February the *Orkney Blast* published this letter ('St Margaret's Monster'):

Royal Artillery, Orkney,
February 11, 1942.

SIR. – After examining the remains of the so called monster now at St. Margaret's Hope, we (undersigned) think the following comments may be of interest to your readers.

The skeleton is entirely cartilaginous; the so-called "blow hole" in the skull communicates directly with the cranial cavity and therefore cannot possibly be a whale type of blow hole. The general make-up of the skull is typically Elasmobranch (sharks, skates, etc.). There is no indication of a true neck in the vertebral column; the first few vertebrae are comparatively small, but they increase progressively in size towards the centre of the body, from which point they decrease towards the tail. The articulation of the first few vertebrae does not differ from that of the rest of the vertebral column, consequently the so-called neck of the animal is no more mobile than any other part of the same.

The pectoral fin contains a large number of radials attached to the typically shark basal plates. The presence of a large number of radials is a non-reptilian characteristic; the number of radials in the fore limb of even in the earlier prehistoric reptiles was, as far as is known, less than in this specimen.

Absence of Jaws

The so-called "hair" was examined and found to be very fine needle-like fin rays formed of cartilage. The tail was incomplete, no indication of a "cow-like" tail was observed.

The stomach was about 2½ feet long [45 cm] and when split open longitudinally contained a spiral valve characteristic of sharks. The contents were not examined, these being badly lacerated.

The absence of jaws and gill supports is of no great significance and may be explained by the fact that this part of the skeleton in sharks is only very loosely attached to the skull and vertebral column. No trace of the upper jaw was seen; in reptiles this is fixed firmly to the skull.

We have little doubt that this specimen, and the skull of another present at St. Margaret's Hope, are the remains of sharks, probably Basking Sharks. –
Yours etc.

J. W. JONES, B.Sc., Ph.D.
W. THORPE-CATTON, M.Sc.

It was not possible to determine if Campbell Brodie had indeed dispatched parts of th skeleton for examination. Instead *The Orcadian* reported on 19 February ('More about th 'Monsters''):

RETURNED TO THE SEA

Because of the public nuisance, caused by the smell from the skeleton, the remains of the Hunda "monster" have been removed from the Red Cross

Society exhibition space near the Post Office, and deposited below high water mark on the village beach at St. Margaret's Hope.

The skeleton may still be inspected by interested parties at low tide. A well known Rear Admiral came ashore one day recently and inspected the skeleton with interest.

News about the Hunda carcase came to an end at that point, so the fate of the skeleton is uncertain, as it seems that it was not preserved and was instead left to the sea.

After the finding of the photograph of the Deepdale Holm Monster, the author came into contact with Claire Jaycock. A week before Christmas in 1941, her great-grandfather James Anderson found the carcase of the '*Scapasaurus*' about 200 yards from his house ('Fresh light on the mystery of the Holm shore 'Monster''). Setting on the track of both cases, she found a picture of another basking shark carcase in the photographic collections of the Orkney Library and Archive and obtained a copy of it (Fig. 4). The photograph was archived under the letters 'St', probably for 'St Margaret's Hope', and on the reverse side the date was given as '1942' (Claire Jaycock, pers. comm., 1 October 2012).

Fig. 4: Photograph of 'Scapasaurus #2' from Hunda (Unknown)

The photograph is in general a good-quality copy, with only slight distortions visible. The image shows the skeleton at the harbour of St Margaret's Hope. The current harbour master, Nick Lovick, located the precise site at the eastern end of Front Road and provided some pictures (Figs 5, 6, 7) and the coordinates (58.826345, -2.956696) of the site (Nick Lovick, pers. comm., 20 April 2012). The skeleton is laid on a wooden board and mounted on two barrels of unknown dimensions. Possibly these are 44-gal (200-l) drums as was believed according to recollections from a contemporary witness (quoted in Nick Lovick, pers. comm., 20 April 2012).

If we assume a standard 44-gal (200-l) drum with a height of 33.5 in (85.1 cm), and a length of the skeleton approximately eight to ten times that length (four to five times visible and - after comparing actual photos of the site - about the same number behind the wall), the skeleton's total length would be between 22 ft (6.7 m) and 28 ft (8.5 m). The dimensions of such a drum would also correspond with the other measurements given for the skull and the 'antennae'.

Figs 5, 6, 7: Current pictures of the site (during high tide) where the skeleton was mounted at St Margaret's Harbour (Nick Lovick)

The overall shape of the vertebral column is clearly that of a shark. The vertebrae are roun(ed) and have no spinous or transverse processes.

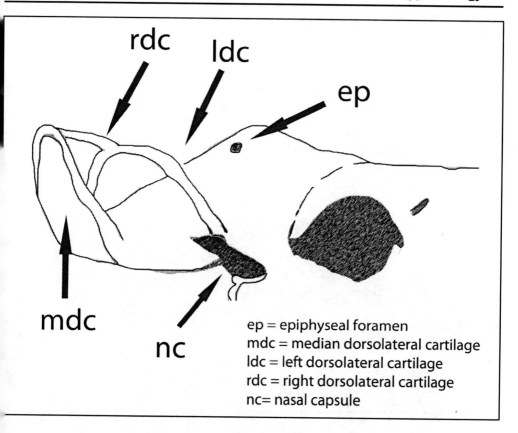

Fig. 8: Part of the skull of *C. maximus* after a picture published in the *Illustrated London News*. The complete skeleton was probably exhibited at the Natural History Museum, London (Markus Hemmler)

Instead, over almost the whole length of the vertebral column the neural arches (through which the spinal cord runs) that are dorsally attached to each vertebra can be seen. In particular, in approximately the middle of the vertebral column the basiventral cartilage can be observed, which makes each vertebra appear triangular in shape. Rotting tissue partially covers them in the photograph, making it impossible to count the vertebrae.

On top of the skull between the eye sockets and housed in a turret-like prominence, the supposed 'blow hole' can be seen. However, it is not the upward-directed blow hole of a cetacean; instead it is an anteriorly-directed opening present in the cranial roof of some elasmobranchs (sharks and rays). This epiphyseal foramen (Fig. 8) is located in the median line of the cranial roof between the anterior margins of the supraorbital crests, and leads to the pineal gland (Hamlett, 1999; de Lulius *et. al.*, 2007). This characteristic foramen of (basking) sharks can be found in various other pseudoplesiosaur cases:

1. Found in September 1808 at Rothiesholm Head, Stronsay, Orkney Islands, Scotland. Sir Everard Home (1809) identified the carcase as a basking shark and most authors favour this opinion (e.g. Bland and Swinney, 1978). Nonetheless, the enormous reported total length of 55 ft (16.8 m) for it has led to the question of whether the *exact* identification as a basking shark is correct (Heuvelmans, 1968; Simpson, 2001; Shuker, 2003). One or two holes on the top of the head had been reported (Oudemans, 2007) and from the position in the drawing of Barclay (1811) we can assume the anterior hole to be the epiphyseal foramen (Fig. 9).

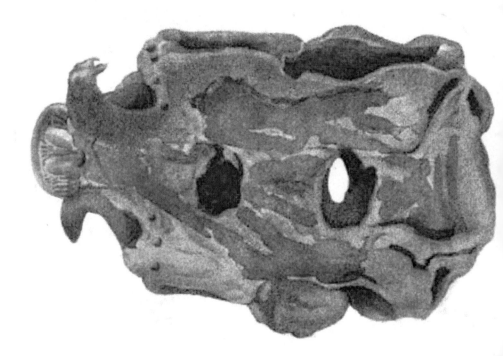

Fig. 9: Skull of the Stronsa carcase (John Barclay)

2. Found in March 1934 at Querqueville, Normandy, France. After inspecting the carcase in situ, the remains were identified by Dr George Petit as *C. maximus* (Heuvelmans, 1968). A short television report from 1934 was included in the documentary film 'Cryptopuzzle' (2001). The carcase was filmed at the beach, and in some close-up shots of the neurocranium the epiphyseal foramen can be clearly seen.

3. Found in January 1939 at Provincetown, Massachusetts, USA. The skeleton was sent to ichthyologist William C. Schroeder who identified it as: "shark, without question a basking shark". While comparing it with another basking shark of 1809, Schroeder also noted: "This latter specimen [of 1809], like the Provincetown one [of 1939], had a hole in the top of its skull" ('Serpent Becomes Basking Shark'; Schroeder, 1939). Again this describes the epiphyseal foramen and examination of the accompanying photograph (Schroeder, 1939) confirms it.

4. Found in December 1941 at Deepdale Holm, Mainland, Orkney Islands, Scotland. Identified as a basking shark by Dr A.C. Stephen (Marwick, 1942b). James Marwick reported:

    > "Head rounded on top with very large eye sockets on either side [...] Down the center of the skull was a hole which I took to be for breathing purposes" (Marwick, 1942a).

    However, it was not a breathing hole but rather the epiphyseal foramen (Fig. 10).

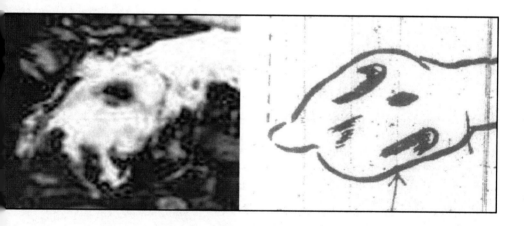

Fig. 10: Photograph and sketch of the Deepdale Holm carcase with "hole for breathing purposes" in the middle of the head between the eye sockets (William S. Thomson; James Marwick)

.   Found in December 1947 at Effingham Inlet, Vancouver Island, British Columbia, Canada. Identified as *C. maximus* by Dr Albert L. Tester from its overall appearance and the presence of some gill rakers and teeth ('Experts Decide 'Sea Serpent' is Only a Shark').

The picture was taken obliquely from the left side of the cranium but the epiphyseal foramen can be readily observed (Fig. 11).

Fig. 11: Skull of the Effingham Inlet carcase (en.wikipedia.org/wiki/File:Effingham_carcase)

6. Found in June 1996 at Block Island, Rhode Island, USA. Before an examination could take place, the skeleton was said to have been stolen. Therefore the identification as *C. maximus* by biologists Lisa Natanson and Harold Pratt was based on photographs ('Tiny Block Island home monster mystery'). The epiphyseal foramen can be seen very clearly (Hamilton, 2008).

7. Found in December 1996 at Claveria, Burias Island, Masbate, Philippines. Identified by ichthyologist Victor Springer as *C. maximus*, this was the first record of a basking shark from the Philippines (Compagno, *et. al.*, 2005). A photograph has been published in *The Philippines Star* but could not be obtained for examination. However, "Its head apparently possessed a blowhole-like orifice, resembling that of a dolphin" (Shuker 2003), which therefore describes the epiphyseal foramen. Again, the author has shown that the descriptive trend of referring to a 'hole' or a 'blowhole' and its identification as the epiphyseal foramen of a basking shark is a helpful indication when assessing 'sea monster' carcases (Hemmler, 2010).

8. Found in June 2004 at Block Island, Rhode Island, USA. After examination of the remains, scientists James Collie and William Macy acknowledged that it was a shark most likely *C. maximus* (Voskamp, 2004). What is evidently a picture of it - due to the location and the form of the skeleton obviously showing the second Block Island carcase - is available online (Esqueleto, 2008), and the epiphyseal foramen is very prominently visible.

Another feature of the supposed photograph of the Hunda carcase is remarkable. To support the snout of a basking shark, the neurocranium has three forward-extending rostral elements on the left and right the rod-like dorsolateral rostral cartilages, and in the middle the much broader median rostral cartilage (Fig. 8). The left and right dorsolateral rostral cartilages aris at the neurocranium in front of the nasal capsules (Izawa and Shibata, 1993; Fairfax, 1998; Hamlett, 1999). During the growth of the shark, the snout and therefore the dorsolateral rostral cartilages alter their shape. Only at a later stage in an adult shark do the dorsolateral rostral cartilages lose their attachment to the tip of the median rostral cartilage and become merely "two prongs" (Heuvelmans, 1968; Fairfax, 1998).

The fact that the dorsolateral rostral cartilages lose their attachment either whilst the shark was growing or by some external circumstances after death has led to the description of 'antennae', 'feelers', 'finger-like projections', or 'horns' in pseudoplesiosaur cases. Four examples of this are as follows:

1. After the identification of the Deepdale Holm carcase, police officer P. Sutherland Graeme, who first reported the corpse, described "two bristle-like projections, one on each side of the upper lip of the animal, each 5.5 inches [14 cm] or more in length and not less than 0.5 inch [1.27 cm] diameter at the base, formed apparently of a substance resembling gristle", and he asked if anyone could explain how they fit into a basking shark identity. "These features had disappeared before Provost Marwick made his careful examination" (Graeme, 1942). Dinsdale (1966) suspected that this could be the 'dorso-rostral' cartilages, which is most probably correct.

2. Dr Bernard Heuvelmans was told that in February 1951 at Hendaye, France, a carcase with "a tortoise's head with two cartilaginous antennae" had been found. Later he was able to collect pictures of the carcase, and identified the dead animal as a basking shark with the 'antennae' as rostral cartilages (Heuvelmans, 1968).

3. Dr Karl P.N. Shuker reported that in April 1998 a carcase was found at Greatstone, Kent, England. The skull was said to "bear a pair of short curved horns". Later Dr Shuker obtained newspaper articles containing photographs of the carcase, and he verified that the 'horns' were actually the rostral cartilages of a basking shark (Shuker, 2003).

4. In September 2002, a pseudoplesiosaur washed ashore at Parkers Cove, Nova Scotia, Canada. Pierre Jerlström and Henry de Roos investigated this case and supported a basking shark identity. While describing the skull, they noticed that "there were also two 150 mm [6 in] finger-like cartilaginous projections, one above each 'nare'" and included a picture of the "head from above showing two finger-like projections" (Jerlström and de Roos, 2005). The described positions above the nares in combination with the picture enables them to be identified as left and right dorsolateral rostral cartilages.

There are certain descriptive trends in reports of mystery carcases that can be helpful for identification purposes (Roesch, 2007), and therefore it was common to assume that the dorsolateral rostral cartilages are the only explanation for such described projections in pseudoplesiosaurs. With the picture (Fig. 4) examined in this paper, however, another option is revealed. As explained before, the left and right dorsolateral rostral cartilages arise at the neurocraniumin in front of the nasal capsules (Izawa and Shibata, 1993; Fairfax, 1998; Hamlett, 1999). The '*antennae*' of the St Margaret's Hope skeleton, conversely, are not located directly in front of the nasal capsules but actually below and much further forward to them (Fig. 12). So they are instead a part of the outer border of the median rostral cartilage, whereas the dorsolateral rostral cartilages have been lost. Possibly only a dark 'knob' from the left rostral cartilage remained.

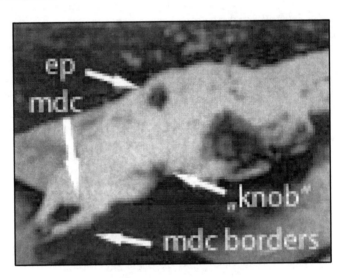

Fig. 12: Skull of the Hunda 'Scapasaurus' with marked features that are mentioned in the text (edited by Markus Hemmler)

Finally: while the author was in contact with harbour master Nick Lovick, he obtained further information by questioning a contemporary witness whom he knew. The photograph in question was shown to this man, who acknowledged that the skeleton portrayed in it was brought to the village and placed near the steps (Nick Lovick, pers. comm., 20 April 2012). Moreover, additional enquiries by Billy Norquay, temporary assistant of Nick Lovick, revealed that his mother, Ms Norquay, can also confirm this from her own experience (Billy Norquay, pers. comm., 10 October 2012).

CONCLUSIONS

The year of the discovered photograph was given as 1942 on its reverse side. The exact location shown in this same photograph could be identified and reveals that the skeleton was mounted at the harbour of St Margaret's Hope. Although not all information in the analysis by Jones and Thorpe-Catton can be revealed, the morphological details fit the descriptions of both scientists as well as showing the skeleton to be from a basking shark *C. maximus*.

In summary, this evidence strongly suggests that this photograph does indeed show the carcase of 'Scapasaurus #2'. The final proof was provided from the contemporary witness kindly contacted by Nick Lovick.

On the basis of this skeleton, the epiphyseal foramen housed in a turret-like prominence of the cranium of *C. maximus* was shown to be responsible for a descriptive trend of holes' or 'blow holes' in pseudoplesiosaurs. Consequently it could be a helpful indication for a possible identification of a basking shark from images and descriptions.

Furthermore the descriptive trend of referring to 'antennae', 'horns', etc, and the identification of these as left and right dorsolateral rostral cartilages, now has to be extended. For as shown here, there is also the possibility that these descriptions are based upon the destroyed or decayed borders of the median dorsolateral cartilage, which should be considered in relation to future cases and to unsolved cases from the past.

ACKNOWLEDGEMENTS
My deepest thanks to archivist Sarah McLean from the Orkney Library and Archive, Markus Bühler, Dr Karl P.N. Shuker, Glen Vaudrey, my nameless contact at *The Orcadian*, the nameless contemporary witness and Ms Norquay of St Margaret's Hope, and above all others especially to Billy Norquay, Nick Lovick, and Claire Jaycock.

Despite considerable efforts we have not been able to trace all rights holders to some copyrighted material. The author welcomes communications from copyrights holders, so that the appropriate acknowledgements can be made in future, and to settle other permission-related matters.

REFERENCES

- (1939) Serpent becomes basking shark. *Provincetown Advocate* (Provincetown), 26 January: 4.
- (1942). More about the "Monster". *The Orcadian* (Kirkwall), 5 February: 3.
- (1942). Fresh light on the mystery of the Holm shore "Monster". *The Orcadian* (Kirkwall), 5 February: 5.
- (1942). Basking shark or Scapasaurus? *Orkney Blast* (Kirkwall), 6 February.
- (1942). The Orkney Monster - First pictures. *Orkney Blast* (Kirkwall), 6 February: 5.
- (1942). Second 'Monster' located. *The Orcadian* (Kirkwall), 12 February: 4.
- (1942). Orkney Monster - Artist's impression. *Orkney Blast* (Kirkwall), 13 February.
- (1942). More about the "Monsters". *The Orcadian* (Kirkwall), 19 February: 4.
- (1942). St Margaret's Monster. *Orkney Blast* (Kirkwall), 20 February.
- (1947). Sea serpent 'investigated'. *Register-Guard* (Eugene), 9 December: 8.

- - (1947). Experts decide 'sea serpent' is only a shark. *Daily Tribune* (Chicago), 15 December: 31.
- - (1996) Tiny Block Island home monster mystery. *Record Journal*. 3 September: 2.
- - (2009) Was ist dranan 'Nessi', 'Yeti' & Co? http://www.bild.de/news/mystery-themen/yeti/schneemensch-ziegenaussauger-lindwurm-9526662.bild.html 27 August. Accessed 28 August 2009.
- Barclay, J. (1811). Remarks on some parts of the animal that was cast ashore on th Island of Stronsa, September 1808. *Memoirs of the Wernerian Natural History Society*, 1: 418-444.
- Bland, K.P. and Swinney, G.N. (1978). Basking shark: genera *Halsydrus* Neill and *Scapasaurus* Marwick as synonyms of *Cetorhinus* Blainville. *Journal of Natural History*, 12(2): 133-135.
- Brodie, W.C. (1942). South Isles "Monster". *The Orcadian* (Kirkwall), 17 February.
- Cohen, D. (1989). *Encyclopedia of Monsters*. Guild Publishing (London).
- Compagno, L.J.V. *et al.* (2005) Checklist of Philippine Chondrichthyes. *CSIRO Marine Laboratories Report 243*.
- 'Cryptopuzzle' (2010). 'Arte', No. 23. Television programme, screened October.
- de Lulius, G., Puilerà, D., and Brown, G. (2007). *The Dissection of Vertebrates - A Laboratory Manual*. Academic Press Elsevier (New York).
- Dinsdale, T. (1966). *The Leviathans*. Routledge and Kegan Paul (London).
- *Esqueleto* [Photograph] (2008). http://marcianitosverdes.haaan.com/wp-content/uploads/2008/06/esqueleto.jpg
- Fairfax, D. (1998). *The Basking Shark in Scotland*. Tuckwell Press (East Linton).
- Graeme, P.S (1942). Query on Holm "Monster". *The Orcadian* (Kirkwall), 5 March: 3.
- Hamilton, S.L. (2008). *Creatures of the Abyss*. Abdo Publishing (Edina).
- Hamlett, W.C. (ed.) (1999). *Sharks, Skates, and Rays: The Biology of Elasmobranch Fishes*. The Johns Hopkins University Press (Baltimore).
- Hemmler, M. (2010). The Masbate Monster. http://www.kryptozoologie-online.de/dracontologie/salzwasserkryptide/masbate-monster-kadaver.html 31 October.
- Heuvelmans, B. (1968). *In the Wake of the Sea-Serpents*. Hill and Wang (New York).
- Home, E. (1809). An anatomical account of the *Squalus maximus* (of Linnaeus), which in the structure of its stomach forms an intermediate link in the gradation of animals between the whale tribe and cartilaginous fishes. *Philosophical Transactions of the Royal Society of London*, 99: 206-220.
- Izawa, K. and Shibata, T. (1993). A young basking shark, *Cetorhinus maximus*, from Japan. *Japanese Journal of Ichthyology*, 40(2): 237-245.
- Jerlström, P. and de Roos, H. (2005). Parkie: a new 'pseudoplesiosaur' washed up on the Nova Scotia coast. *Journal of Creation*, 19(2): 109-118.
- Marwick, J. (1942a). A strange creature. *The Orcadian* (Kirkwall), 29 January: 5.
- Marwick, J. (1942b). Not anything interesting! *The Orcadian* (Kirkwall), 19 February: 4.

- Oudemans, A.C. (2007). *The Great Sea-Serpent* (reprint). Coachwhip Publications (Landisville).
- Roesch, B. (2007). A lesson about tusked sea-serpent carcases. http://web.archive.org/web/20011016083058/http://www.forteantimes.com/exclusive/roesch_01.shtml. Accessed 2 October 2012.
- Schroeder, W.C. (1939). The Provincetown "sea serpent". *New England Naturalist*, No. 2: 1-2.
- Shuker, K.P.N. (2003). *The Beasts That Hide From Man*. Paraview Press (New York).
- Voskamp, P. (2004). The return of Block Nessie. http://www.blockislandtimes.com/News/2004/07/03/Front_Page/005.html 3 July. Accessed 5 September 2004.

IDENTIFYING 'JAWS', THE MARGARET RIVER MAMMAL CARCASE

Darren Naish
Ocean and Earth Science,
National Oceanography Centre,
University of Southampton,
Southampton SO14 3ZH, UK.
Email: eotyrannus@gmail.com

ABSTRACT

A partially decomposed mammal carcase, discovered and photographed on a sandy beach in the Margaret River area of Western Australia in 1975, and nicknamed 'Jaws', has been mooted as an unidentified enigma and never satisfactorily identified in print. On several occasions, it has been linked with proposals that modern Australia might still be inhabited by thylacoleonids ('marsupial lions'). The carcase also has to be viewed within the context of proposals that large, non-native felids and atypically large feral domestic cats might exist in the country. Partial decomposition, poor photographic resolution, and the absence of any unambiguous scale make it difficult to comment without ambiguity on many aspects of the carcase's morphology. Nevertheless, the details of its dentition are clear, as are its gross overall anatomy and shape. Based predominantly on the configuration and form of the teeth, the carcase can be identified as that of a carnivoran, specifically a felid. It almost certainly is a domestic cat *Felis catus*.

Keywords: mammal – carcase – Australia – carnivoran – marsupial – Margaret River

INTRODUCTION

The alleged presence of mysterious, cat-like carnivorous mammals on the Australian mainland remains an area of interest and controversy. Eyewitness descriptions, tracks, livestock kills, and photographic evidence suggest that dark and tan-coloured large cats – most resembling puma *Puma concolor* and leopard *Panthera pardus* – are present (at least occasionally) as feral species in the Australian outback (O'Reilly, 1981; Shuker, 1989; Healy and Cropper, 1994), and at least three large felid specimens (the St Arnaud puma of 1924, the Woodend puma of the 1960s, and the Broken Hill lioness of 1985) were shot dead in Victoria and New South Wales during the 20th century (Williams and Lang, 2010). Whether such animals represent odd and occasional escapees from private and/or public collections or whether they represent members of persistent and breeding populations remains undetermined.

Similarly, the alleged presence of especially large feral domestic cats *Felis catus* in parts of

Australia is of uncertain significance. A black cat, reportedly at least 1.6 m long in total length, was shot by Gippsland-based hunter Kurt Engel in 2005. The body was discarded but DNA tests of surviving tissue apparently identified the animal as a specimen of *F. catus* (Williams and Lang, 2010). Video footage (e.g. that obtained by Gail and Wayne Pound at Lithgow, New South Wales, in 2001) seems to show feral domestic cats with shoulder heights of c. 60 cm living wild in parts of Australia. The exceptional size of these feral domestic cats remains to be verified; their presence could well explain at least some 'big cat' sightings.

While the presence of both non-native, escapee big cats and unusually large feral domestic cats could both explain the Australian big cat phenomenon, it has repeatedly been suggested in the grey, popular, and semi-popular literature that some sightings of large cat-like Australian mammals are not of cats in the true sense, but perhaps of extant members of Thylacoleonidae, a group of predatory marsupials often popularly referred to as 'marsupial lions'. The fossil record indicates that the youngest known thylacoleonid taxon – *Thylacoleo carnifex* – did not persist any more recently than c. 46,000 years ago, and it seems to have gone extinct at about the same time as the rest of the Australian megafauna (Murray, 1984; Roberts et al.; 2001, Wroe and Field, 2001). Far younger dates of c. 7500 years ago, obtained via early attempts at radiocarbon dating (Daily, 1960), have not been verified by more recent work, and are at odds with radiocarbon estimates and faunal correlation dating the youngest remains of this species to c. 46,000 years ago (Dawson, 1985; Roberts et al., 2001). Rock art that depicts a large, vertically striped, superficially cat-like mammal has been interpreted by some authors as evidence that *Thylacoleo* was seen, and its presence recorded, by humans (Akerman, 1998; Akerman and Willing, 2009; Taçon et al., 2011).

Nevertheless, based on eyewitness descriptions of a long-tailed, stripy, leopard-sized mammal supposedly seen in Queensland and elsewhere (and popularly known as the 'Queensland tiger') Heuvelmans (1995) and others (Shuker, 1989, 1995; Healy and Cropper, 1994) intimated or suggested that thylacoleonids might persist to the present in Australia. However, despite the early popularity of this proposal, it failed to win adherents in the technical zoological community and has never been supported by material evidence or photographic documentation. Despite this, the thylacoleonid proposal has persisted, with the hypothesised identification now having shifted to the plainer, dark- or tan-coloured 'big cats' reported by Australian eyewitnesses. This idea is based on the fact that some eyewitness reports describe peculiar animals that differed from true cats in detail (see Healy and Cropper, 1994; Williams and Lang, 2010), but, again, it lacks material support and can only be regarded as extremely speculative, albeit highly intriguing.

It is within the context of the supposed present-day persistence of thylacoleonids that the piece of evidence discussed here was first brought to light. During or around 1975, a group of

---

Fig. 1: Photograph showing 'Jaws' carcase, photographed on beach in the Margaret River area, West Australia, with close-up of the animal's head below. No unambiguous scale is present. The surrounding sand grains and pieces of seaweed create the approximate impression that the carcase is less than 100 cm in total length.

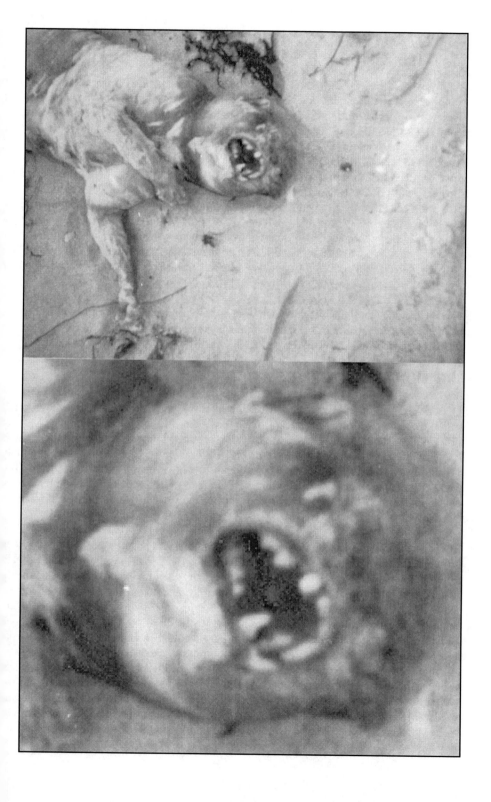

beachcombers in the Margaret River area in Western Australia discovered and photographed a partially decomposed mammal corpse (Fig. 1). A cropped version of the single known colour photograph was published in Karl Shuker's 1996 book *The Unexplained* (I am not aware of any other published versions of the image). Shuker obtained the photo from Australian correspondent Kevin Farley; Farley noted that staff at both the Australian Museum and the Queensland Museum had been unable to identify it (Shuker, 1996). As discussed below, such pronouncements are not necessarily all that meaningful. Due to kind co-operation from Jeff Johnson, I was able to obtain digital scans of the photo in 2009, viewable at larger resolution than the version reproduced in Shuker (1996). Interested in resolving the identity of the carcase, I hoped to compare the visible morphological details of the carcase (especially the dentition) with those of other mammals and hence arrive at a conclusion. At the time of writing, I am not aware of any published effort to do likewise. However, during my time on the editorial board of the (now defunct) *The Cryptozoology Review*, I was aware of a submitted manuscript in which an author attempted to identify those teeth visible in the photos and arrive at an identification of the animal based on this interpretation (B. Speers-Roesch, pers. comm., 2002). I have never seen this manuscript and its conclusions remain unknown to me.

It must be noted at the outset that the unknown whereabouts of the carcase (or of any of its bones), combined with the poor quality of the surviving photograph, means that we will likely never be able to arrive at a conclusive identification.

DESCRIPTION AND INTERPRETATION

The photo shows a partially furred mammal carcase, lying on its left side on a sandy substrate (Fig. 1). Seaweed and other pieces of detritus are in close association; based on the inferred size of these objects the carcase appears to be less than 100 cm in total length. The left forelimb is lying extended away from the body and against the substrate while the right is folded against the chest and upper region of the left forelimb. The head is tilted obliquely upwards and away from the substrate. The mouth is open, revealing most of its teeth. The tongue and/or palate are not visible. Hindlimbs, a tail, and indeed the entire pelvic region are not included in the photos but the animal's body shape, approximate proportions, and forelimb shape appear typical for a quadruped.

Few details of the carcase can be described without ambiguity, but it is partially covered in pale, off-white fur across much of its body and forelimbs (note that the figures included here depict the photo in black and white, not in its original, colour version). Some of this fur appears to be present across the underside of the lower jaw and dorsal surface of the snout and face. However, the fur is patchily distributed, with expanses of brown and orange-tinged naked skin being exposed across the abdomen, part of the chest, and upper arms. The skin exposed on the neck and head is a deeper reddish-purple colour. It is assumed that this partial hairlessness is due to the early onset of decomposition of the carcase in water: after weeks in water, mammal carcases typically lose some, most, or all of their hair, with a naked-skinned carcase being the eventual result (pers. obs.). The upper arms appear well muscled, and the proximal part of the lower arm is well muscled and bulbous on its ventral surface, indicating the presence of large flexor muscles. The wrist region is gracile (being less than half the dorsoventral thickness of the proximal section of the limb), and the extremities of the limbs

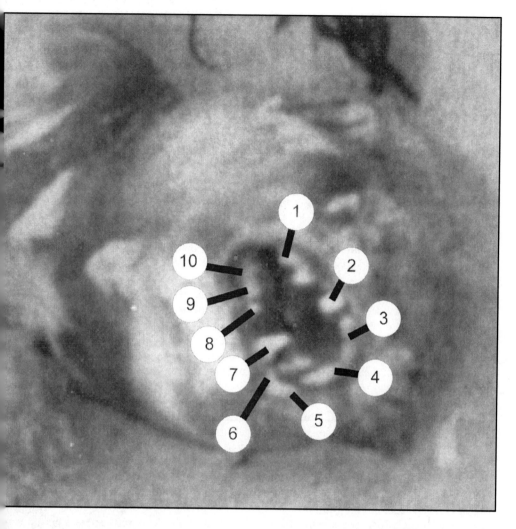

Fig. 2: Face and dentition of the Margaret River carcase, with teeth labelled (see text for corresponding description of numbered teeth).

while not readily visible in detail due to low resolution – appear to bear short, blunt-tipped digits that form paws. There is no indication of any opposable digit, and the animal's right paw possesses at least three subparallel digits that strongly recall those of carnivorans like cats and dogs.

The head is proportionally wide and short-snouted, with a muscular, convex temporal region. Structures that look like the right eye and ear can be discerned but do not reveal any useful information. A dark area close to the mid-line of the snout may represent a displaced and distorted rhinarium. Patches of fur are present on the underside of the lower jaw and apparently across some or all of the left side of the face and snout. A dark, slim lip band lines the edges of the mouth. The teeth are white and appear in good condition; there are no obvious indications that the jaws or teeth are displaced (due to fracturing or distortion of the skull) relative to the condition in life (Fig. 1).

The form and configuration of the dentition can be described in reasonable detail. The numbers listed here correspond to those included in Fig. 2. At the very back of the right side of the upper jaw, two, closely spaced teeth are present, both of which sport centrally placed, tall, subtriangular cusps (1). The right upper canine is clearly shorter and blunter than the left (2). Its crown is evidently broken. Note that it is separated by a short gap from the two subtriangular-cusped teeth just mentioned. It is not possible to say whether this gap was present in life, or whether it was originally occupied by a small tooth. Small upper incisors are evidently present (3). Cropped versions of the photo create the impression that two large incisors were present, but close examination of larger images shows without doubt that a larger number of small incisors are actually present. The left upper canine is complete, unlike the right (4). The tooth is slim, sub-conical, gently curved, and with a pointed tip. The left lower canine is similar in shape to the left upper canine, but definitely shorter. The lower canines are shorter than the uppers in virtually all mammals. A few small white specks in the photo, located close to the canines, are interpreted here as small incisors (6). While it is not possible to be sure about this identification, the objects are in the right place to be lower incisors and also match them in size, shape, and proportions. Only two or three are visible; they are less than half the height of the lower canines. The right lower canine is conical (7). A substantial gap separates the right lower canine from the first post-canine tooth (8). The gap is about equivalent to half the length of the lower canine. The first post-canine appears short (both anteroposteriorly and in terms of cusp height) and simple, but no detail is visible. The second post-canine is much longer anteroposteriorly, and also taller in terms of cusp height (9). No detail is really visible, but it seems to have two cusps at least. The third post-canine is the largest and tallest of the three (10). There is the suggestion of a tall anterior cusp, followed by a notch. There might be a tall posterior cusp further back.

The obvious presence in the carcase of small upper and lower incisors, large upper and lower canines, and an assortment of post-canine teeth (Fig. 2) immediately expunges several mammal groups from the list of possible contenders, most notably diprotodontian marsupials (including possums, macropods and koalas), all of which lack lower canines and possess only two, hypertrophied lower incisors (Archer et al., 1999; Horovitz and Sánchez-Villagra, 2003). Any suggestion that the carcase might be that of an extant thylacoleonid can therefore be

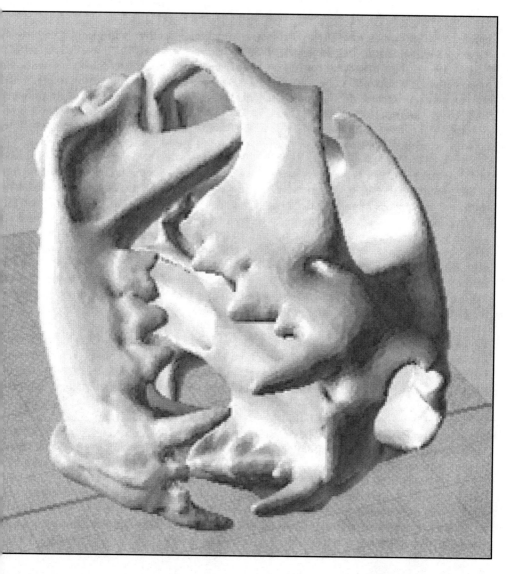

Fig. 3: Digital model of a domestic cat skull tilted on its side to show form and onfiguration of the dentition. Compare with Figs 1 and 2. Image by Colin McHenry.

ignored; thylacoleonids have highly distinctive, protruding incisors and gigantic carnassials (Owen, 1859; Anderson, 1929; Pledge, 1975; Archer and Dawson, 1982; Wells *et al.*, 1982 Nedin, 1991). The observed configuration of teeth appears most consistent with the placental clade Carnivora. Other groups that have been informally suggested to match the carcase (e.g., Primates) do not match the carcase as well: monkeys and other primates with large canines, for example, have low-crowned (rather than tall-crowned) post-canines, larger, more prominent incisors, and a thorax that is generally wide and shallow rather than deep and narrow as seems to be the case in the Margaret River carcase.

In carnivorans, the eye is typically positioned immediately above the most posterior teeth in the toothrow (that is, a straight vertical line can be drawn between the most posterior teeth and the posterior part of the eye) (Knight, 1947; Antón *et al.*, 1998). When this observation is extended to the Margaret River carcase, it appears that the most posterior visible teeth are probably the most posterior teeth in the dentition (though the possible presence of one more posterior tooth does remain). In the upper jaw, two post-canine teeth can be seen. The space immediately posterior to the canine allows the possibility of a third tooth, and the fact that there might be one additional tooth right at the back of the toothrow (and out of view) allows the possibility of a fourth tooth. At absolute maximum the carcase therefore had four upper post-canines.

This conclusion immediately rules out dogs (where there are always five post-canines and essentially only leaves cats and mustelids. Mustelids can probably be ruled out from further consideration for several reasons: mustelid carcases generally appear longer headed and longer-snouted than the Margaret River carcase, and their limbs are typically shorter and with longer, more clearly differentiated digits. Furthermore, the idea that the carcase is that of a cat is obviously far more parsimonious than other, far more speculative possibilities (e.g. the carcase represents a new mustelid taxon). A felid identification is in line with the other details of the carcase: it is, after all, a short-faced animal with large, slim canines and gracile limbs. Its near-parallel upper canines, gently diverging lower canines, and neat, parallel incisors also look exactly cat-like: in short faced dog breeds, the canines diverge strongly, and the incisors typically project at odd angles, looking decidedly uneven, and are separated by irregular gaps. Comparison with an image of a rotated small felid skull (Fig. 3) reveals an identical configuration of similarly-shaped teeth.

DISCUSSION

If the Margaret River carcase is that of a felid, the obvious question is whether it is that of a domestic cat *Felis catus*. In the frustrating absence of size data, it remains impossible to be completely sure, but – as noted above – the impression created by the apparent size of the surrounding sand and seaweed is that this is not a particularly big animal. Given that domestic cats (feral or otherwise) occur globally, are tremendously abundant, and are well known component of the Australian ecosystem, and in the absence of any contradictory evidence from the animal's overall appearance or morphological details feral domestic cat clearly represents the most sensible identification.

Given that cats are familiar animals, and that the details of their anatomy are (in contrast to those of so many animals) accessible and well documented in easily available literature, it might seem peculiar that experts at two major museums were supposedly unable to identify the Margaret River carcase (Shuker, 1996). While the exact details of the specific interactions are unknown to me and, in this case, unreported in the literature, previous experience with similar cases suggests the possibility that the relevant experts were misquoted (a statement to the effect that a definite identification cannot be reached becomes interpreted as synonymous with "It's unidentifiable"), or that no relevant experts ever really examined the evidence in question (there are lots of experts at museums, but not all are experts at identifying mammal corpses).

The mystery surrounding the Margaret River 'Jaws' carcase can now be regarded as resolved, and the existence of this specimen does not have any bearing on the purported persistence of thylacoleonid marsupials beyond the end of the Pleistocene and into modern times. While the exact size of the carcase cannot be determined, there are no indications that it was anything other than a normal-sized domestic cat. As such, the carcase does not have any special relevance for the hypothesis that non-native felid species, or that atypically large feral domestic cats, are abroad in Australia.

ACKNOWLEDGEMENTS

I thank Jeff Johnson for sending scanned versions of the 'Jaws' carcase, and Karl Shuker and Ben Speers-Roesch for discussion. Two reviewers provided comments which helped improve the manuscript. Colin McHenry kindly created and provided the image used here as Fig. 3. Comparison with my late pet cat, Tigger Mamun-Ra, assisted research. The hypothesis (and some of the text) presented here originally appeared on the author's blog, Tetrapod Zoology, in May 2009.

REFERENCES

Akerman, K. (1998). A rock painting, possibly of the now extinct marsupial *Thylacoleo* (marsupial lion), from the north Kimberley, Western Australia. *The Beagle, Records of the Museum and Art Gallery of the Northern Territory*, 14: 117-121.

Akerman, K. and Willing, T. (2009). An ancient rock painting of a marsupial lion, *Thylacoleo carnifex*, from the Kimberley, Western Australia. *Antiquity*, 83(319). http://www.antiquity.ac.uk/projgall/akerman319/

Anderson, C. (1929). Palaeontological notes no. I: *Macropus titan* Owen and *Thylacoleo carnifex* Owen. *Records of the Australian Museum*, 17: 36-49.

Antón, M., García-Perea, R. and Turner, A. (1998). Reconstructed facial appearance of the sabretoothed felid *Smilodon*. *Zoological Journal of the Linnean Society*, 124: 369-386.

Archer, M., Arena, R. *et al.* (1999). The evolutionary history and diversity of Australian mammals. *Australian Mammalogy*, 21: 1-45.

- Archer, M. and Dawson, L. (1982). Revision of marsupial lions of the genus *Thylacoleo* Gervaise (Thylacoleonidae: Marsupialia) and thylacoleonid evolution in the late Cainozoic. *In:* Archer, M. (ed.), *Carnivorous Marsupials*. Royal Zoological Society of New South Wales (Sydney): 477-494.
- Daily, B. (1960). *Thylacoleo*, the extinct marsupial lion. *Australian Museum Magazine*, 13: 163-166.
- Dawson, L. (1985). Marsupial fossils from Wellington Caves, New South Wales; the historic and scientific significance of the collections in the Australian Museum, Sydney. *Records of the Australian Museum*, 37: 55-69.
- Healy, T. and Cropper, P. (1994). *Out of the Shadows: Mystery Animals of Australia*. Ironbark (Chippendale).
- Heuvelmans, B. (1995). *On the Track of Unknown Animals* (updated edit.). Kegan Paul International (London).
- Horovitz, I. and Sánchez-Villagra, M.R. (2003). A morphological analysis of marsupial mammal higher-level phylogenetic relationships. *Cladistics*, 19: 181-212.
- Knight, C.R. (1959). *Animal Drawing: Anatomy and Action for Artists*. Dover Publications (New York).
- Murray, P.F. (1984). Extinctions down under: a bestiary of extinct Australian Late Pleistocene monotremes and marsupials. *In*: Martin, P.S. and Klein, R.G. (eds), *Quaternary Extinctions: A Prehistoric Revolution*. University of Arizona Press (Tucson): pp. 600-628.
- Nedin, C. (1991). The dietary niche of the extinct Australian marsupial lion: *Thylacoleo carnifex* Owen. *Lethaia*, 24: 115-118.
- O'Reilly, D. (1981). *Savage Shadow: The Search for the Australian Cougar*. Creative Research (Perth).
- Owen, R. (1859). On the fossil mammals of Australia. Part II. Description of an almost entire skull of the *Thylacoleo carnifex*, Owen, from a freshwater deposit, Darling Downs Queensland. *Philosophical Transactions of the Royal Society*, 149: 309-322.
- Pledge, N.S. (1975). A new species of *Thylacoleo* (Marsupialia: Thylacoleonidae), with notes on the occurrence and distribution of Thylacoleonidae in South Australia. *Records of the South Australia Museum*, 16: 261-267.
- Roberts, R.G., Flannery, T.F. *et al.* (2001). New ages for the last Australian megafauna: continent-wide extinction about 46,000 years ago. *Science*, 292: 1888-1892.
- Shuker, K.P.N. (1989). *Mystery Cats of the World: From Blue Tigers to Exmoor Beasts*. Robert Hale (London).
- Shuker, K.P.N. (1995). *In Search of Prehistoric Survivors: Do Giant 'Extinct' Creatures Still Exist?* Blandford (London).
- Shuker, K.P.N. (1996). *The Unexplained: An Illustrated Guide to the World's Natural and Paranormal Mysteries*. Carlton (London).
- Taçon, P.S.C., Brennan, W. and Lamilami, R. (2011). Changing perspectives in Australian archaeology, part XI. Rare and curious thylacine depictions from

Wollemi National Park, New South Wales and Arnhem Land, Northern Territory. *Technical Reports of the Australian Museum, Online*, 23(11): 165-174.

- Wells, R.T., Horton, D.R. and Rogers, P. (1982). *Thylacoleo carnifex* Owen (Thylacoleonidae): marsupial carnivore? *In:* Archer, M. (ed.), *Carnivorous Marsupials*. Royal Zoological Society of New South Wales (Sydney): 573-586.
- Williams, M. and Lang, R. (2010). *Australian Big Cats: An Unnatural History of Panthers*. Strange Nation (Hazelbrook).
- Wroe, S. and Field, J. (2001). Mystery of megafaunal extinctions remains. *Australasian Science*, 22(8): 21-25.

INSTRUCTIONS TO CONTRIBUTORS

These fall into two categories: important issues to consider when preparing a cryptozoological paper; and the style of presentation required for submissions to the journal.

(i) Important Issues to Consider When Preparing a Cryptozoological Paper

The *Journal of Cryptozoology* aims to publish papers of equal rigour to existing mainstream science journals. Cryptozoology is a controversial topic. Consequently, the following guidelines suggest important issues to consider when preparing a cryptozoological paper, in order to pre-empt common criticisms that might be made by reviewers.

Premises and Background

1   The scientific literature works from the existing consensus. By definition (i.e. Heuvelmans, 1988), there is no consensus on whether the putative animals behind cryptozoological reports exist. Therefore papers submitted to the *Journal of Cryptozoology* should *not* start from a premise that a specific animal species does exist (except to make predictions about the evidence for it that might be found in future or to develop methodology). Papers can of course present evidence for the existence of unknown animals, and argue against the existing consensus.

2   Papers should work from the existing scientific consensus with regard to other arguments as well. This does not mean non-consensual assumptions cannot be made but that if such assumptions are made, they should be justified by reference to evidence. For example, if an author believes that certain bipedal tracks are made by a putatively still-extant prehistoric animal that palaeontologists assume is quadrupedal (and extinct), then some evidence for the presumed animal's bipedality, *independent* of the tracks, should be given.

Professions of personal belief in particular cryptids are not part of a scientific paper, but personal experience of seeing putatively unknown animals may be suitable for publication.

It is particularly important that authors should *not* presume the zoological identity of a particular cryptid (see also Point 17 below), nor should language be employed that implies the identity is unequivocally known. Papers can of course argue for a particular zoological explanation for a certain set of reports, given particular evidence and test hypotheses concerning particular identities.

5   Authors should be aware of and cite the relevant scholarly literature on the topic in question. The use of *Google Scholar* or similar search engines allows basic searching of the scholarly literature by people outside of research institutions. References used should be scholarly where possible, and definitely from print media or an online equivalent. Ephemeral references should not be used except as an absolute last resort. Non-scientific blogs and similar articles are generally unacceptable as sources in formal papers, as is *Wikipedia*.

6   Authors should be aware that words can have a technical meaning within certain fields that is different from that in general use. For example, the terms 'accuracy' and 'precision' have specific technical meanings in statistics, although the words are often considered synonymous by the general public. Likewise, 'bug' as a strict zoological term is restricted to one particular taxonomic order of insects, as opposed to its more general use in North America for almost any terrestrial arthropod.

Using Eyewitness Testimony

Unlike conventional zoology, cryptozoology often uses eyewitness testimony as a potential source of evidence. Authors should be aware of the numerous biases associated with eyewitness testimony in general (e.g. Loftus, 1996) and in a specifically cryptozoological framework (e.g. Arment, 2004; Paxton, 2009).

7   Given witnesses cannot, by definition, know what it is that they are reporting, they cannot necessarily gauge their own accuracy or precision.

8   Authors should not necessarily assume eyewitnesses are correct in their identification of body parts and/or the taxonomic affinities of what they are reporting. In addition, there may also be ambiguity in the eyewitness reports. For example, is a reported 'mane' a lion-like mane or a horse-like mane or something different from both of these?

9   Just because reports come from a given area does not mean they have the same origin.

10   Just because reports are hypothesised to have the zoological source (i.e. cryptid) does not mean they actually have the same source. For instance, a given report of a bigfoot could be of an unknown animal, of a man in a gorilla suit, or of a variety of known animal. The origin of reports superficially of the same type need not be consistent in space.

11   Catalogued eyewitness reports are an outcome of a long process that includes *acquisition* (i.e. the perception of the original event), *retention* (i.e. memory), *retrieval* (i.e. recollection), transmission, and recording (Loftus, 1996; Paxton, 2009). At each and every stage, biases may creep in, which means that accessible reports may represent a very inaccurate and imprecise sample of what was actually seen.

12   From Point 11, it is very possible that collected cryptozoological reports suffer from sampling biases that affect the conclusions drawn from them. For example: if aquatic cryptids are reported most in calm weather conditions (e.g. Mackal, 1976), does this reflect the biology

of an unknown animal or is it due to increased detectability under those conditions? (See Buckland *et al.*, 2001 for a discussion of detectability in an animal survey context.)

13   The most frequent source of bias in cryptozoological reports is that they record presence only. If conclusions want to be drawn from reported occurrences or report frequencies, then this can only really be done if search effort (e.g. man-hours searching, area searched, effort in collecting reports, etc) is known.

14   It follows from Point 13 that raw quantification of reports without taking account of search effort *cannot* be used to infer putative habitat preferences of cryptids, etc.

15   Trends in features of reports, if real, may reflect cultural changes of biases in the reporting process rather than changes in the biology of the putative animals, if any.

16   Wherever possible, the original wording of reportees should be used. Authors should avoid interpolating meaning to witness statements unless absolutely necessary.

17   Cryptids are hypothetical constructs of what the putative source of eyewitness reports might be. The raw data of eyewitness-based cryptozoology is reports. Data and hypotheses should not be confused. It is therefore inappropriate to state "bigfoot occurs in the forested areas of British Columbia", when what is incontestably true is "bigfoot reports come from the forested regions of British Columbia".

Reaching Conclusions

All conclusions in the *Journal of Cryptozoology* should be evidence-based - where the premises, data, and chain of reasoning leading to the conclusion are clearly stated. Every point made *must* be justified either through evidence and argument supplied within the paper or by reference to existing published literature. Care should be made in distinguishing between possible explanations of things, of which there are an infinite number, and probable explanations of things, of which there are few and which are of rather more interest to the readership. Probable rival hypotheses for explanations of certain phenomena should be explicitly stated and the reasons for rejecting/accepting a particular hypothesis over others explicitly given.

See also Arment (2004) and Paxton (2011) for more commentary on the methodology of cryptozoology.

References

Arment, C. (2004). *Cryptozoology, Science and Speculation*. Coachwhip (Landisville).
Buckland, S.T., *et al.* (2001). *An Introduction to Distance Sampling*. Oxford University Press (Oxford).

- Heuvelmans, B. (1988.) The sources and methods of cryptozoological research. *Cryptozoology*, 7: 1-21.
- Loftus, E.F. (1996). *Eyewitness Testimony*. Harvard University Press (Cambridge).
- Mackal, R.P. (1976). *The Monsters of Loch Ness*. Futura (London).
- Paxton, C.G.M. (2009). The plural of "anecdote" can be "data": statistical analysis of viewing distances in reports of unidentified giant marine animals 1758–2000. *Journal of Zoology*, 279: 381–387.
- Paxton, C.G.M. (2011). Putting the "ology" into cryptozoology. *Biofortean Notes*, 1: 7-20.

(ii) The Style of Presentation Required for Submissions to the Journal

All submissions must be original manuscripts not previously published elsewhere or submitted elsewhere simultaneously with submission to this journal. All submissions will be sent to two members of the journal's peer review panel for their opinions concerning content, clarity, and relevance to cryptozoology. Their comments will then be studied by the editor whose decision is final concerning whether or not the manuscript is published, subject if necessary to amendments by the author(s) if suggested by the reviewers. The copyright of all published papers belongs to this journal.

All manuscripts submitted should be one of the following three types of paper:

Discussion/Review article:

Its subject should be a discussion or literature review of a given cryptozoological subject, and should not include original, unpublished research. It can be of 1000-4000 words in length, and can also include clearly labelled and numbered b/w photographs, artwork, tables, or maps provided that the copyright of these falls into one of the following three categories:

(1) owned by the author(s);

(2) has been granted to them in writing by their copyright owner(s) - a copy of such permission will need to be submitted with the manuscript and artwork;

(3) expired, i.e. in the public domain.

The article should be preceded by a 200-word abstract, and should be divided into relevant subtitled sections. A reference list can be included at the end of the article; if so, this and the accompanying in-text citation style should correspond with the preferred version outlined below.

Research article:

Its subject should be original research (but not fieldwork) conducted by the author(s). It should be of comparable length to or shorter than discussion/review articles, but with a minimum count of 1000 words. It can also include clearly labelled and numbered b/w photographs, artwork, tables, or maps, provided that the copyright of these falls into one of the three above-listed categories. The article should be preceded by a 100-200 word abstract, and its main text should be split into four sections – Introduction, Materials and Methods (or Description where more appropriate), Results/Interpretation, Discussion/Conclusions. A reference list can be included at the end of the article; if so, this and the accompanying in-text citation style should correspond with the preferred version outlined below.

Field report:

Its subject should be fieldwork conducted by the author(s). It should be of 1000-2500 words in length. It can also include clearly labelled and numbered b/w photographs, artwork, tables, or maps, provided that the copyright of these falls into one of the three above-listed categories. The article should be preceded by a 200-word abstract, and its main text should be split into four sections – Introduction, Description (in which the fieldwork undertaken is described), Results, Discussion (which should also include details of any future plans). A reference list can be included at the end of the article; if so, this and the accompanying in-text citation style should correspond with the preferred version outlined below.

Style of reference citation required:

All in-text citations should be: author(s) surnames, comma, year of publication, all in parentheses. If the cited reference has more than two co-authors, give only the first surname followed by *et al.* Examples: (Jones, 1987), or (Jones and Jones, 1987), or (Jones *et al.*, 1987).

For books, the style required for the reference list should be: Author surname followed by given names as initials, then followed by the year of publication in parentheses, and a full stop/period. The title of the book should be italicised, with its principal words beginning with a capital letter, and should end with a full stop/period. The publisher's name should then be given, with the town or city of publication included in parentheses. If the book is co-authored by two authors, their names should be separated by 'and'; if co-authored by more than three, only the first author's name should be given, followed by a comma and then '*et al.*' (in italics). Here are some hypothetical examples:

Smith, J.C. (1987). *The History of Cryptozoology.* Jones and Son (London).
Smith, J.C. and Jones, J.A. (1987). *The History of Cryptozoology.* Jones and Son (London).
Smith, J. C., *et al.* (1987). *The History of Cryptozoology.* Jones and Son (London).

For journal articles, the style required for the reference list should be: Author surname followed by given names as initials, then followed by the year of publication in parentheses, and a full stop/period. The title of the article should not be italicised, and should not be capitalised (other than for the first word or proper nouns). The title of the journal should be given in full, not abbreviated, with its principal words beginning with a capital letter, it should be italicised, and should end with a comma. Volume numbers should be given as figures, issue numbers also as figures (preceded by no.) but included in parentheses following the volume number (together with date of issue if relevant, and separated from issue number by a semi-colon), followed by a colon, and then the page numbers, given in full. If the article is in a newspaper, the town or city of publication in parentheses should follow the newspaper's title, and instead of volume numbers, the full date of publication will suffice, followed by the page number(s) if known. Here are some hypothetical examples:

- Smith, J. C. (1987). Investigation of an unidentified lizard carcase discovered in Senegal. *Journal of Lizard Studies*, 33 (no. 2; September): 52-59.
- Smith, J. C. (1987). Mystery cat on the loose in Wales. *Daily Exclusive* (London), 4 February: 23.

For online sources, if an author name is given, it should be presented in the same style as for books and articles, followed by the title of the source, which should adhere to the style format given above for a hard-copy journal article, followed by the complete URL, date of posting if given, and the date upon which it was accessed by the paper's author(s). Here is an example:

- Shuker, K.P.N. (2012). Quest for the kondlo – Zululand's forgotten mystery bird. http://www.karlshuker.blogspot.com/2012/02/quest-for-kondlo-zululands-forgotten.html 21 February.

If no author is given, simply begin the reference with - , then give the article title, etc as above.

References inserted directly in the paper's main text should take the form of: Smith (1988) Jones (1989). Or, if cited within brackets in the main text, they should take the form of (Smith 1988; Jones, 1989).

A brief note on illustrations

Do not embed illustrations into MS Word documents. Please provide them each as separate attachments; monochrome images at a minimum of 100 dpi, colour images at a minimum of 250 dpi. If there is a technical reason why this cannot be done, please email publisher@journalofcryptozoology.com

Submissions

Please email all submissions to the Editor at editor@journalofcryptozoology.com

CPSIA information can be obtained at www.ICGtesting.com
Printed in the USA
BVOW08s2236191213

339680BV00002B/4/P